Thrust and drag

Thrust is the force that moves an object or a person. **Friction** is the force that stops things moving. In a racing bike the thrust is produced by the rider's legs. Friction occurs between the tyres and the track. There is also air resistance or **drag**, which works against the thrust. The effects of drag can be reduced by **streamlining**. This means making sure that the air flows smoothly over the object.

▶ In games like rugby and American football the ball can be made to go further by making sure it is kicked or thrown so it travels point first. This means there is only a little air resistance or drag. The same idea is used when racing cars are designed.

Here are other examples of thrust, drag and streamlining.

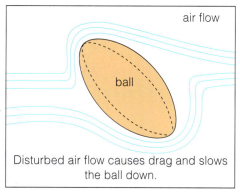

Disturbed air flow causes drag and slows the ball down.

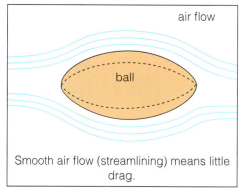

Smooth air flow (streamlining) means little drag.

Q1 For each of the three photographs, write down where these effects occur:
a thrust;
b drag;
c streamlining.

Q2 In the experiment that your teacher demonstrated what gave the ball
a thrust forces?
b drag forces?

Extension exercise 2 can be used now.

Sprint starts and thrust

Apparatus

☐ personal scales (in newtons)
☐ block of wood

You are going to find out if you can get more thrust using a **sprint** start or a standing start. You will need to work with a partner.

Q1 Copy this table.

Position	Force (in newtons)	
	Left leg	Right leg
Standing		
Sprint		

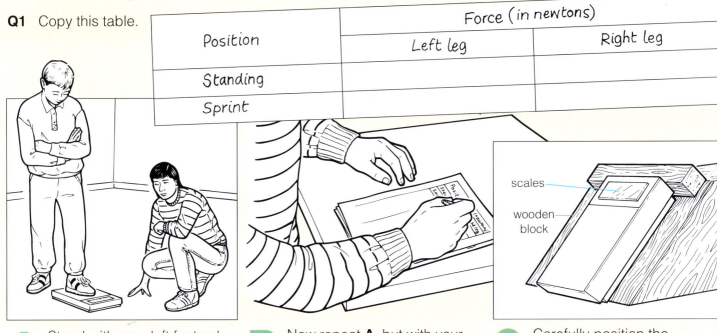

A Stand with your left foot only on the scales. Without moving your right foot, push down as hard as you can with the left. *Do not jump onto the scales.* Record your result in the table. ▲

B Now repeat **A**, but with your right foot on the scales. Record your result. Now your partner must do **A** and **B**. ▲

C Carefully position the wooden block in the angle of the wall and floor. Rest the scales against the block. Make sure it is steady. ▲

scales
wooden block

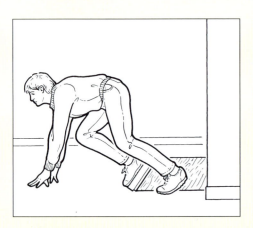

D Now take up the sprint start position (as shown in the diagram) with your right foot on the scales. Push backwards as hard as you can. Ask your partner to read the result. Write it down in the table. Then repeat the test with your left foot on the scales. ◀

E Next your partner must carry out **C** and **D** and write down the results. ▶

Q2 Collect the results from the class and draw a bar chart to shown them.

Force (N)
400
300
200
100

Left leg   Right leg

Q3 Which start position gave you the most thrust?

Q4 Which leg was the stronger?

Q5 Which leg would you 'start on' in a race?

1 What has science got to do with sport?

Q1 For each photograph write down one way in which you think science may have helped this sport.

2 Forces

Throwing angles

Your teacher will demonstrate this experiment.

Q1 Predict what will happen to the distance the plasticine travels as the angle of release gets bigger.

Q2 Copy this table.

Your teacher will:

Angle of release (degrees)	10	15	20	25	30	35	40	45	50	55
Distance ball travels (cm)										

A Use the apparatus carefully and make sure that the ball of Plasticine is always round. ▼

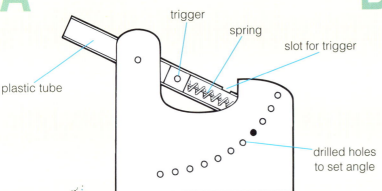

trigger
spring
slot for trigger
plastic tube
drilled holes to set angle

B Set the angle of the tube to 10 degrees. Put the ball of Plasticine into the tube and fire. Measure the distance the ball travels. ▼

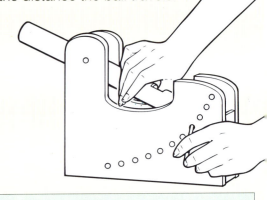

C Record the first distance in your table. Then repeat **B** for the other angles. Fill in your results table. ▼

Q3 Was your prediction correct?

Q4 At which angle did the ball travel furthest?

Q5 Plot a line graph of your results. Copy the axes shown below.

Q6 What else, other than angle, may affect your results?

Q7 Design and carry out a similar experiment to see if the shape of a **mass** affects how far it goes.

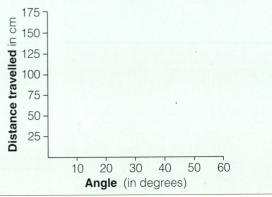

Distance travelled in cm
175
150
125
100
75
50
25
Angle (in degrees)
10 20 30 40 50 60

Extension exercise 1 can be used now.

Footwear, surfaces and friction

The force that lets the sole of your shoe grip the floor when you are running is called friction. Sports footwear has to give suitable friction with the surface on which it is used. The surface must also give suitable friction.

There are three main types of sports footwear:
☐ Studded boots – football, rugby and other field sports.
☐ Spikes – athletics.
☐ Tennis/training shoes – indoor sports/hard surfaces *e.g.* badminton, tennis.

These types of footwear are used to give the right grip on different surfaces. ▶

Some types of surface

Grass	Allows different kinds of footwear. Gives poor friction when wet or frozen.
Tartan track	**Synthetic** surface, not affected by wet conditions. Provides good surface grip. Grip is improved by use of spikes.
Wood	Good friction, except when wet.
Astroturf	All weather surface. Provides good surface for grip. Causes 'burns' when the skin is rubbed on it.

Running shoes and trainers

Here are some hints on buying shoes or trainers.
☐ Avoid long shoelaces. You might trip over them.
☐ There should be about 1 cm between your toes and the end of the shoe. This gives your feet space to expand when they get hot.
☐ Avoid plastic shoes. Your feet may get too hot.
☐ Avoid shoes with heel tabs. They may rub your ankles.
☐ Get shoes with thick, cushioned soles to protect your feet.

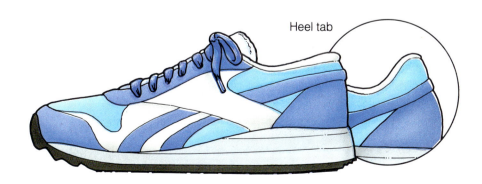

Heel tab

Q1 Hazel has been chosen to run the 100 metre sprint for the county. She has been given some money for some new footwear. Which type of footwear do you think she should buy? Why?

Testing grip

You are going to find out how different soles and studs affect the grip of sports shoes on different surfaces.

Q1 Copy this table.

Type of sport shoe	Force needed to move shoe (in newtons)		
	Carpet/grass	Sand/soil	Hard surface

Apparatus

☐ selection of shoes such as:
football boots
athletics spikes
running shoes
plimsolls
multi-stud shoes
☐ 1-kg mass
☐ 0-50 N **newtonmeter**
☐ damp sand/soil tray
☐ piece of carpet or grass.

A First place the football boot on the carpet or grass. Put the 1-kg mass inside it. Attach the newtonmeter to the boot. ▲

B Now pull the newtonmeter gently towards you. Watch the scale on the meter carefully. Read the force used to just move the boot. Record the result in your table. ▲

C Now repeat **A** and **B**, still using the boot, but first on the sand/soil and then on a hard surface. Record your results. ▲

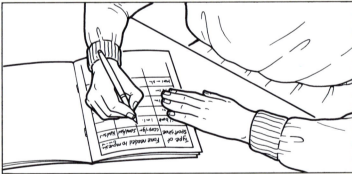

D Next repeat **A**, **B** and **C** using the other types of sports shoes. Record your results. ▲

Q2 Which shoe was the easiest to move on:
a the hard surface
b the carpet/grass
c the sand/soil?

Q3 Why do footballers have different studs for wet and dry conditions?

Q4 Which was the best footwear for gripping
a the hard surface
b the carpet/grass
c the sand/soil?

Q5 Using your results explain why there are different types of sports shoes.

Winter sports

Friction is important in winter sports. Ice and snow have two
properties important to sport:
a low friction
b they melt under pressure

The fact that ice and snow melt under pressure is made use of in
skating and skiing.

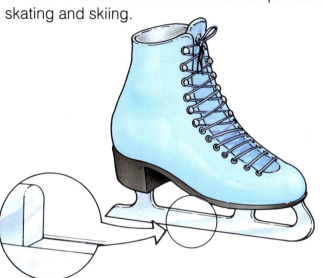

▲ The thin edge of a skate blade produces great
pressure. This pressure melts the ice and means that
the skating actually takes place on a thin film of water.
This gives extremely low friction.

▲ The thrust of a skier is given by the sticks or poles.
The friction between the snow and the skis can be
reduced by rubbing wax on the underside of the skis.

The effect of pressure on ice

You are going to see what happens
to ice when pressure is put on it.
Your teacher will do this experiment
as a demonstration.

The apparatus will be set up as in
the diagram. Look at it carefully at
the beginning of the experiment.
Look at it again every five minutes.

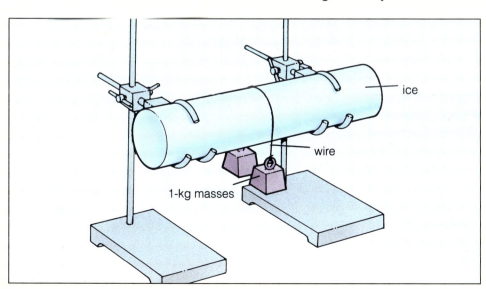

Apparatus

☐ cylinder of ice
☐ 30 cm of thin wire
☐ two 1-kg masses
☐ clamps and stands

Q1 What happens to the ice
below the wire?

Q2 What happens to the wire as
it passes through the ice?

Q3 What would happen if you
increased the pressure of the
wire on the ice?

Gripping ice

You are going to watch an experiment to find out which surfaces give a good grip (high friction) and which give poor grip (low friction). Remember, in winter sports low friction is most important.

Apparatus

☐ ice cubes
☐ 1 waxed plank (1 metre)
☐ 1 unwaxed plank (1 metre)
☐ stop clock ☐ ruler
☐ wooden ladder

Q1 Copy this table.

Your teacher will:

Surface	Height at which ice moved (cm)	Time taken (seconds)
Unwaxed wood		
Waxed wood		

A Rest the unwaxed wood on the first step of the ladder. Next put one of the ice cubes on the unwaxed wood. As the teacher lets go of the ice cube, start the stop clock. ▲

B If the mass slides, time how long it takes to reach the bottom. If it does not slide keep raising the wood until the ice cube does move. Record the height at which the ice cube moved. ▲

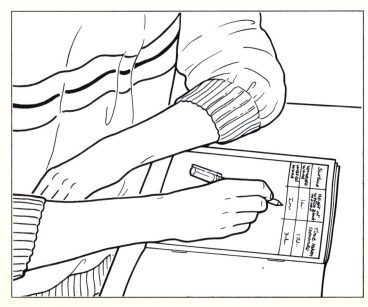

C Record your results in the table. Your teacher will then repeat **A** and **B** using the waxed wood. Write down your results. ◄

Q2 Which surface needed the least height before the ice cube moved?

Q3 Which surface gave the least friction with the ice cube?

Q4 Why do skiers wax their skis?

Q5 Brian has an old sledge with metal runners. What advice would you give him to make it go faster?

Does the surface of a ball affect its speed?

Golf balls travel through the air very quickly. In this experiment you are going to use a gel to slow golf balls down. You will then be able to time their drop and see if their different surfaces affect the speed they move at.

Apparatus

☐ two golf-ball-sized pieces of baked clay or Plasticine, one with dimples, one without
☐ 1-litre measuring cylinder containing gel
☐ stop clock (digital)

Q1 Copy this table.

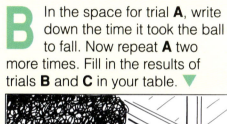

Type of surface	Time of drop (seconds)			Average $= \dfrac{A + B + C}{3}$
	Trial A	Trial B	Trial C	
Dimpled				
Smooth				

A Hold the dimpled ball just at the top of the gel. As you let the ball go, start the stop clock. When the ball reaches the bottom stop the clock. ▼

B In the space for trial **A**, write down the time it took the ball to fall. Now repeat **A** two more times. Fill in the results of trials **B** and **C** in your table. ▼

C Repeat **A** and **B** using the smooth ball. Record your results. Then work out the average time for each ball to drop to the bottom. ▼

Q2 Which ball fell faster and why?

Q3 Why do cricketers shine the ball on one side only?

Q4 Why is a cricket ball changed during a match?

3 Balls in sport

Apparatus

☐ selection of balls (cricket, golf, squash, table tennis, tennis, soccer, rugby, hockey)
☐ electronic balance
☐ calipers
☐ metre ruler

Bounce

You are going to collect data (information) on different types of balls and how they bounce.

Q1 Copy this table, adding spaces for all the balls you will test.

Ball	Diameter (cm)	Mass(g)	Material (e.g. leather, plastic)	Height of bounce		

A Using the calipers and the metre ruler, measure the diameter of each of the balls. Record the results in your table. ▼

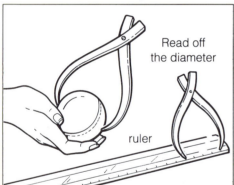

Read off the diameter

ruler

B Now use the balance to find the mass of each ball. Record the results in your table. ▼

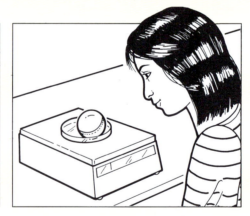

C Next you must inspect each ball carefully and decide what it is made of. Record your results in the table. ▼

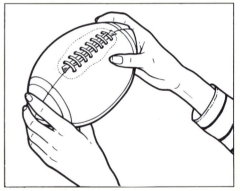

D Now measure the bounce of each ball. Drop the first ball from the top of the metre ruler. Watch carefully to see how high it bounces. ▼

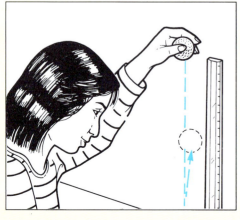

E Record the result. Then repeat **D** two more times. Next repeat **D** three times for each of the other balls. Make sure you drop the ball from the same point each time. ▼

F Record all your results carefully in the table. ▼

Q2 Which ball bounced the highest?

Q3 Did any of the balls bounce back to the height from which they were dropped?

Bounce and elasticity

Bounce is one of the most important features of a ball. The amount of bounce needed in a ball varies from game to game. Some games have a rule about the bounce of the ball. For example, in basketball, a ball dropped from 2 metres onto a hard floor must bounce to between 1.2 metres and 1.4 metres.

▶ For a ball to bounce it must be **elastic**. When a ball is elastic it 'gives' when it is hit or when it hits a hard surface. This is because the ball is usually softer than the surface it hits. Being elastic means that the ball can store **energy** for a short time and then release it as it returns to its original shape.

A ball will never bounce back to the height it fell from. Some of the energy used to squash the ball changes into other forms of energy. The energy is not lost or destroyed. Some of it changes into less useful energy forms.

☐ Some of it changes into heat energy.
☐ Some of it changes into sound energy.
☐ Some of it changes into movement energy and squashes the surface the ball has landed on.

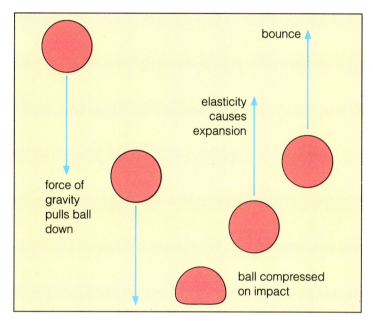

bounce

elasticity causes expansion

force of gravity pulls ball down

ball compressed on impact

Q1 Explain why balls will not bounce back to the same height that they were dropped from.

Q2 Do all balls in sport 'give' when they hit a surface?

Q3 Why does a squash ball get hot when it is hit many times?

Q4 When a cricket ball is dropped into sand, why does it not bounce?

4 Fitness

The five 'S's – stamina, strength, suppleness, speed, skill

▲ **Stamina** is the ability to keep going during a sport. To improve stamina you have to do exercises which make your heart and lungs stronger and able to work harder. Activities that improve stamina include swimming, aerobics classes, walking, jogging and cycling.

▲ **Strength** is the amount of force muscles can produce. Different sports develop different sets of muscles. To develop muscular strength you have to do exercises that use your muscles, such as weight training, running, push-ups and squat thrusts.

◀ **Suppleness** is an important part of being fit. It affects the way you move and means you are less likely to get injured. Regular stretching exercises and warm-ups make you more supple.
Supple people can easily bend, stretch and twist into different positions. As we become older we become less supple.

Continental Sports

Q1 What happens to your body temperature when you do exercises that need strength and stamina?

Q2 Which muscles do weight lifters use? (Think carefully.)

Q3 Which muscles do footballers mainly use?

Q4 Which muscles get stronger as you train doing step-up exercises?

Q5 Which of these people would have
a the most suppleness?
b the most stamina?
sprinter   gymnast   weight lifter   footballer
tiddlywinks player   marathon runner

Speed in sport can be divided into two groups. Athletic speed, or moving the body quickly, and reaction speed. Different sports need different types of speed. Sprinting is mainly athletic speed, while table tennis involves a lot more reaction speed. Both types of speed can only be improved by practice and training.

▶ All sports have their own particular **skills** which have to be learned. Good tennis players are not always good badminton players even though they all use rackets. Good footballers are not always good rugby players, but both games are team games using balls.

Q6 Choose one sport and, using the information on these two pages, list the different skills and qualities you would need to be good at that sport.

Some suppleness exercises to try

A Keeping your legs straight, try to touch your toes. Try this again with your heels against a wall. ▼

B Hold a metre rule across your shoulders and behind your head. Stand up very straight. How far can you turn your body from your waist (*not* your hips)? This may be easier if you stand against a bench. ▼

<table>
<tr><td>**Apparatus**</td></tr>
<tr><td>☐ metre rule</td></tr>
</table>

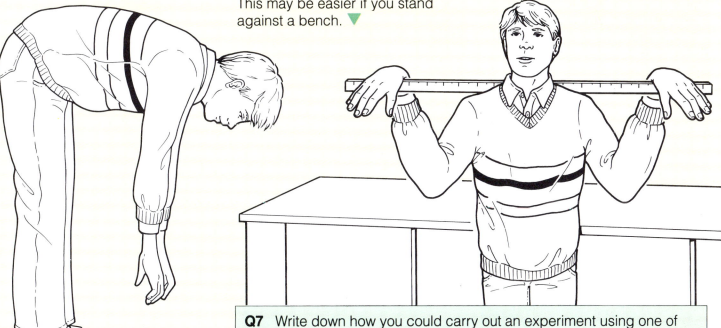

Q7 Write down how you could carry out an experiment using one of these exercises to measure the suppleness of different people in your class. Remember to think about how to measure and record your results and how to carry out the experiment safely. Discuss your plan with your teacher, then try it out.

What is fitness?

▼ These photographs show people who are especially 'fit' for their sport. They may be fit in different ways. They may have great strength. They may be very supple. They may have great speed or quick reactions. They may have great stamina or staying power. As well as these, they also have special skills for their sport.

Marathon running

Climbing

Badminton

Fishing

Gymnastics

Show jumping

Football

Canoeing

Tennis

Q1 Do you think you are fit?

Q2 Explain your answer to Q1.

Q3 Look carefully at each of the photographs in turn. For each picture write down if the person is using strength, suppleness, speed, stamina or any other special skills.

Stamina

Someone with good stamina can keep going when other people get tired and give up. It is one way of measuring your fitness. A person's stamina is linked to the rates at which their heart and lungs work.

During this experiment, you will measure heart rate by taking **pulses**. If you have not done this before ask your teacher for help.

Q1 Copy this table.

1st pulse reading	2nd pulse reading	3rd pulse reading

A Start the clock. Do not stop the clock until you have filled in all of the table. Step on and off the box 30 times per minute for three minutes. ▼

B After the exercise sit down and rest. When the clock reads four minutes get your partner to take your pulse for 30 seconds. Record this in your table. ▼

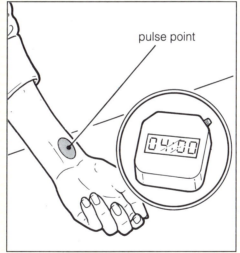

pulse point

C Repeat pulse readings when the clock reads five minutes and six minutes. Record these in your table. ▼

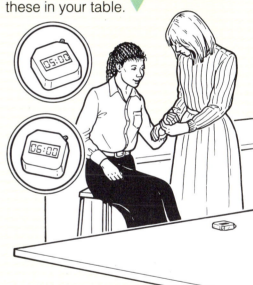

Q2 What happened to your breathing rate while you were resting?

Q3 Now you can work out your heart/lung fitness score. Copy this table and fill in the totals.

Q4 Look at the scores table below. Use the total from **D** to see how you did.

A Add together your 1st, 2nd and 3rd pulse rates.	Total...
B Multiply this total by 2.	Total...
C Now work out 180 seconds × 100.	Total...
D Then divide the number in **C** by the total in **B**.	Total...

Scores	
90 or more	Superior stamina
80 – 89	Excellent
70 – 79	Good
60 – 69	Fair
50 – 59	Poor
less than 50	Oh dear!

Extension exercise 3 can be used now.

Effects of exercise

You are going to plan and carry out an investigation into the effect of exercise on the body.

Answer the following questions. They will help you to think about what you need to do to plan and carry out your investigation.

Q1 Make a list of the changes that take place in your body when you exercise.

Q2 Can all of these changes be measured?

Q3 How can you measure these changes? Write your ideas down.

Q4 What measurements do you need to make before you start exercising?

Q5 How are you going to record your results? A table? A graph?

Q6 Is your investigation a fair test? If not, how can you improve it to make it fair?

Q7 Now write out a rough plan of your investigation. Discuss it with your teacher and then you may be able to carry it out.

Strong heart and lungs

In order for muscles in the body to work, they need **oxygen**. The more work they do the more oxygen they need. Oxygen gets into the body through the **lungs**.

Oxygen travels around the body to the muscles in the blood. Blood is pumped around the body by the heart.

When you are resting, your average heart beat is between 70 and 80 beats per minute. Some very fit sportsmen and sportswomen have resting heart rates as slow as 38 beats per minute.

Their training makes their hearts work harder. Gradually their heart muscles get stronger and bigger. Their hearts need to beat fewer times per minute to pump blood around the body.

▶ These athletes have a heart rate of 38 beats per minute at rest.

▲ Sportsmen and sportswomen also practise breathing skills. Runners learn to take deep breaths before a race to get plenty of oxygen into their blood.

Swimmers have to control their breathing to fit in with their strokes. Many athletes have large lung capacities and strong chest muscles. Lung disease will affect a sportsman or woman's fitness.

Very fast sprinters do not breathe at all during a race and then breathe hard afterwards to clear their lungs.

Q1 When we exercise our hearts beat faster. Why does this happen?

Q2 What happened to your heart rate after the exercise in your stamina test?

Q3 Do you need more oxygen when you are walking or when you are running?

Q4 Why do athletes not smoke?

Q5 Roy was late again for work. He had to run for the bus. When he sat down he felt as if his heart was going to 'burst'. What was actually happening?

Extension exercise 4 can be used now.

Recovery rates

Apparatus

☐ stop clock
☐ wooden block 30 cm high

⚠ Take care when doing this exercise if you have been ill recently or if you have a heart or lung problem. Make sure the box cannot slip during the experiment. Get somebody to hold it steady.

Another measure of fitness is the rate at which the heart and lungs return to their resting rates after exercise, in other words how fast you can recover.

Before doing this experiment you will have to count how many times your heart beats in 30 seconds when resting and how many times you breathe in 30 seconds while resting.

This is easier to measure if you work with a partner.

Q1 Copy this results table. Leave space to put in more results if you need to.

| | Over 30 seconds | |
	Pulse	Breaths
Resting		
After exercise		
+ 1 min		
+ 2 min		
+ 3 min		
+ 4 min		

A Start the stop clock, and keep it going until the experiment is over. Do step-ups for three minutes. ▼

B Get your partner to take your pulse for 30 seconds as soon as the exercise stops. Note how many breaths you take in 30 seconds. ▼

C Repeat **B** every minute until your pulse returns to normal. Do not forget to record the results each time in minutes. ▼

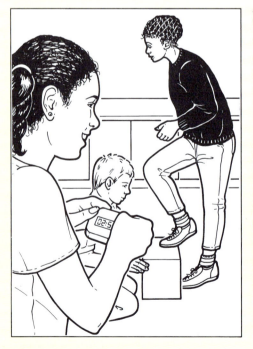

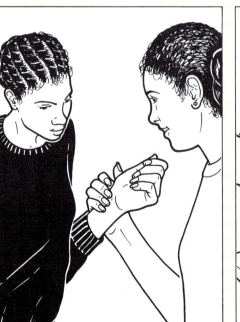

Q2 How long did it take your breathing to return to resting rate?

Q3 How long did it take your pulse rate to return to resting rate?

Q4 Compare your results with the rest of the class. Who is the fittest?

5 Muscles and movement

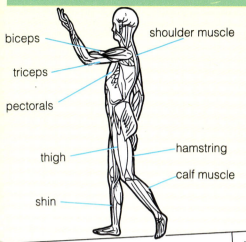

biceps
triceps
pectorals
thigh
shin
shoulder muscle
hamstring
calf muscle

Apparatus

☐ bathroom scales measuring in newtons
☐ mat or cloth for use on the floor

Your muscles

Muscles and bones work together to make your body move. The diagram on the left shows some of the important ones.

You are going to do some investigations to help you identify some of the muscles in your body. You will measure the force they can produce in newtons (N).

Q1 Copy this table.

	Triceps	Biceps	Finger	Pectorals	Thigh
Reading on scales in newtons					

A You are going to carry out **B** to **F**. At each step ask your partner to read the scales and record the results. ▲

B Put the scales on a table. Push your fists firmly down on the scales. The muscles that are working are the **triceps**. ▲

C Put the scales under the edge of a table. Push upwards on the scales, with hands flat. You are now testing your **biceps**. ▲

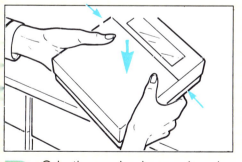

D Grip the scales in your hands. Squeeze. The finger muscles are now working. ▲

E Hold the scales between the palms of your hands. Push them together. You are now working your **pectoral** muscles. ▲

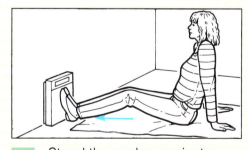

F Stand the scales against a wall. Push the scales with both legs. You are now using your thigh muscles. ▲

Q2 Which set of muscles produced the most force?

Q3 On your body, find all the muscles shown at the top of the page.

Q4 Name a sport which develops strong thigh muscles.

Q5 Name a sport which develops biceps and triceps muscles.

Bones and muscles

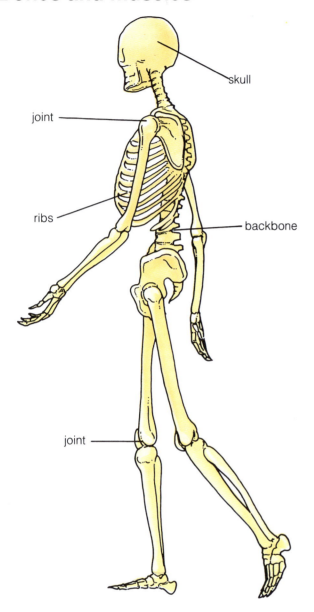

◀ There are over 200 bones in the human body. Together, all your bones make up your **skeleton**. The bones are joined together by **ligaments**. You can only move a **joint** when muscles pull on the bones of your skeleton.

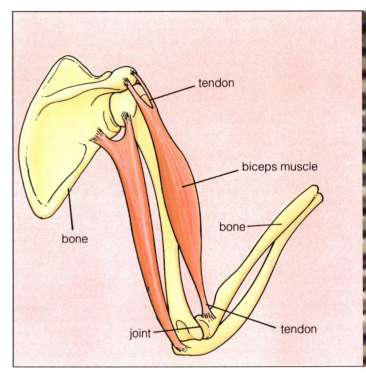

▲ There are over 350 muscles in the human body.

Each muscle is attached to a bone by **tendons**. A joint between bones moves when muscles 'contract' or get shorter. The muscle pulls on the tendon. The tendon pulls on the bone. The bone then moves. Tendons and ligaments are made from tough material. They will not stretch easily.

You are going to search for your Achilles tendon. ◀

A Put your hand on the back of your ankle. Move your foot up and down.

B Feel for bones and muscles. Between the foot bones and the muscle at the back of your leg is a 'cord' or 'string' This is the Achilles tendon.

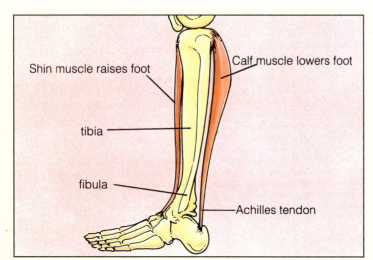

Q1 What does your Achilles tendon do?

Q2 Which muscle pulls on the Achilles tendon?

Working muscles

What some of your muscles do	
Biceps and **triceps** muscles	Raise and lower the arm at the elbow.
Pectoral and **shoulder** muscles	Move the whole arm forwards and backwards.
Thigh and **hamstring** muscles	Bend and straighten the knee joint.
Calf and **shin** muscles	Bend the ankle joint.

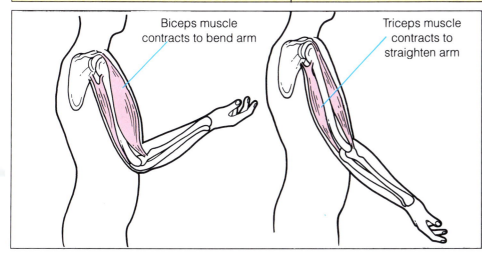

Biceps muscle contracts to bend arm

Triceps muscle contracts to straighten arm

◀ To bend a limb one muscle must pull on two bones. Muscles always work in pairs. One bends or 'flexes' a joint, the other straightens or 'extends' the joint. Such pairs of muscles are called **antagonistic**. This means they do the opposite of each other.

▼ These diagrams show how leg muscles work in the first two strides of a race.

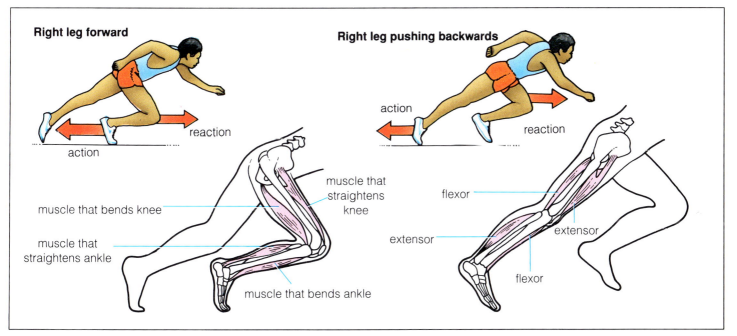

Right leg forward

reaction

action

muscle that straightens knee

muscle that bends knee

muscle that straightens ankle

muscle that bends ankle

Right leg pushing backwards

action

reaction

flexor

extensor

extensor

flexor

Q1 Why do sportsmen and women often eat extra protein when they are in training?

Q2 Apart from food, what else do muscles need when they work? (Look back to page 17 if you cannot remember.)

Q3 In a sprint race, why does a crouch start, or using blocks, give a faster start than standing up straight?

Muscles and work (1)

You are going to measure how much work is done by some of your muscles. You will lift masses of 20 newtons a distance of 1 metre.

Work done = force (in newtons) x distance moved (in metres).

Work is measured in units called **joules**.

To lift one 20 N mass 1 metre takes 20 joules of work.

Q1 Copy this table.

Force of each mass (in newtons)	Height of table (in metres)	Number of masses lifted in 10 seconds	Work done (in joules)
20 N	1		

A Collect some masses of 20 N. Adjust the height of the table or stool to 1 metre using the wooden blocks. ▼

B Start the stop clock and lift the masses one by one as fast as you can onto the table. Do this for 10 seconds. ▼

C Record the number of masses lifted in your results table. Work out the total work done. ▼

Q2 If you repeated this experiment exactly with masses of 40 N would you do more or less work?

Q3 What else were you lifting apart from the 20 N masses?

Q4 Lisa and Jon have had an argument about who works the hardest then they are running. They both run 400 metres in exactly the same time. Lisa weighs 500 N and Jon weighs 600 N. Who do you think is working the hardest? Explain your answer.

Muscles and work (2)

Work (measured in joules) **= Force** (measured in newtons) × **Distance** (measured in metres)

Q1 This athlete has a body weight of about 860 N. He can pole vault up to a height of about 8 metres. How much work do his muscles do to raise him up to that height? ▲

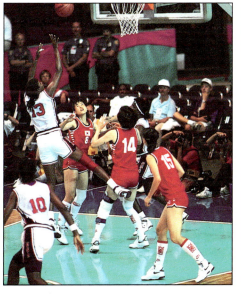

Q2 Weight lifters need a lot of strength in their arm muscles. How much work is done to lift a total weight of 900 N to a height of 2 metres above the ground? ▲

Q3 Basketball players need to be very tall to reach their goal easily. A basketball has a weight of 5 N, and a basketball net is about 3 metres from the ground. How much work is done to lift the ball to the net from a waist height of 1 metre? ◄

Q4 In competitions, why are javelins, shots, discus, cricket balls, *etc.* all an exact weight?

Energy and sports

Everybody needs energy to stay alive. When you make your body work hard you use extra energy. We get our energy from food.

Bowls	1030 kJ
Badminton	1580 kJ
Tennis	1760 kJ
Football	2260 kJ
Squash	3800 kJ
100 metre sprint	7200 kJ

This table shows the amounts of energy, in kilojoules, needed to do different sports for one hour.
1 kilojoule (kJ) = 1000 joules.

A 100 metre sprint usually lasts for about 10 seconds. The sprinter will use 20 kJ of energy.

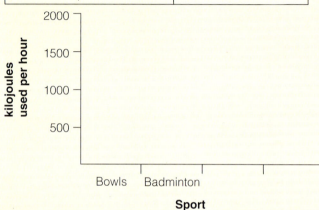

Q1 Which sport uses up more energy; football or badminton?

Q2 If you played squash for half an hour, how much energy would you use?

Q3 Using the information given in the table, draw a bar chart on graph paper. Use axes like the ones shown on the left.

Energy-giving foods

In the experiment on page 26 you will burn muscle fuel. Muscle fuel is some form of sugar.

◀ Energy-giving foods are mostly **carbohydrates**. They include potatoes, rice, pasta, bread, plantain, cake, jam, fizzy drinks, and squash.

Q4 During long tennis matches the players often drink sweet drinks. Explain why.

Q5 Swimmers often take glucose tablets before a race. Explain why. (Glucose is a type of sugar.)

Q6 The day before a long-distance cycle race or a marathon race the athletes eat very large amounts of carbohydrate. Explain why.

Extension exercise 5 can be used now.

Strength and shape

Muscles affect the shape and fitness of your body. By exercising you keep your muscles firm. Firm muscles give the body its shape.

▼ If you work your muscles very hard and train several times a day you could look like the people in these pictures.

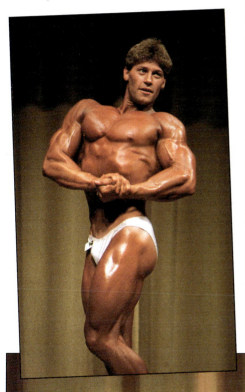

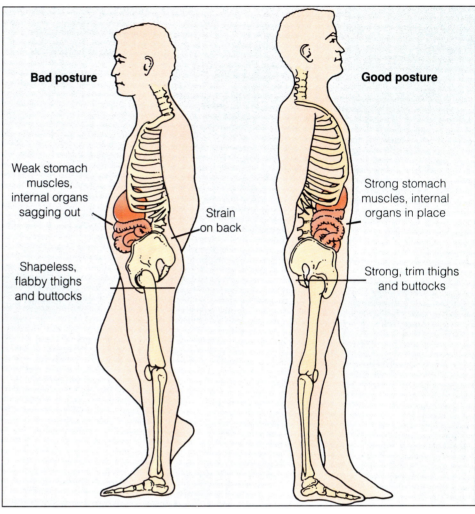

Bad posture

Weak stomach muscles, internal organs sagging out

Strain on back

Shapeless, flabby thighs and buttocks

Good posture

Strong stomach muscles, internal organs in place

Strong, trim thighs and buttocks

Muscles are made of protein. To build up new or bigger muscle you need to exercise your muscle and eat protein foods. Protein foods include meat, fish and eggs.

The more use muscles get, the stronger they become.

Q1 Why do sportsmen and women often eat extra protein when they are in training.

Q2 Apart from food, what else do muscles need if they do a lot of work? (Look back to page 17 if you cannot remember.)

Muscle energy

For work of any kind to be done there has to be a source of energy. The energy that makes your muscles work comes from the food you eat. Energy is released in the muscles by a process called **respiration.** As well as food, the muscles need oxygen to release energy.

You are going to do an experiment to show the amounts of energy given off by muscle fuel.

Q1 Copy this table.

Sample of muscle fuel	Temperature of water at start (°C)	Temperature of water at end (°C)	Temperature change (°C)
1			
2			

A Measure 3 g of muscle fuel into a bottle cap.
Put the cap on a gauze on top of a tripod and Bunsen burner. ▲

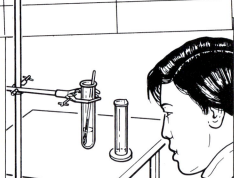

B Now measure 10 cm³ of the water and pour it into the test tube.
Take the temperature of the water in the tube.
Put the test tube into the clamp. ▲

C **Put on your eye protection**. Light the Bunsen burner and heat the sample of muscle fuel very strongly until it burns. **Turn off the Bunsen.** ▲

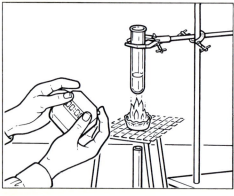

D Place the test tube above the burning sample. **Take care.** Start the stop clock. ▲

E After one minute use the heatproof mat to put out the flames. **Take care**. Record the final temperature of the water in your results table. ▲

F Check you have all the results you need. Then work out the temperature change from the muscle fuel sample. Now repeat **A** to **F**. ▲

Q2 Which sample gave the greatest rise in temperature?

Q3 Why was exactly the *same* amount of *cold* water used for both experiments?

Q4 Why might your results not have been very accurate?

6 Centre of mass

Centre of mass and balance

The **centre of mass** is the point on an object where gravity acts. For evenly shaped objects like a football, the centre of mass is in the centre. For most objects, for example humans and rackets, the position of the centre of mass is not obvious as they are not evenly shaped.

Anything can be balanced if its centre of mass is directly above its base. The lower the centre of mass, the more likely it is to stay above its base.

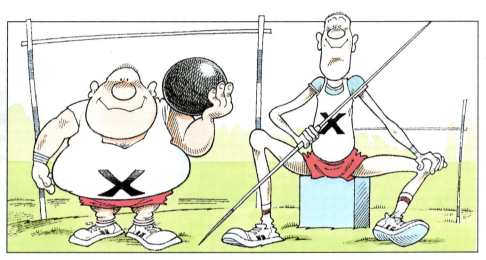

◀ A large base makes an object more **stable**. Short fat people are more stable than tall thin people.

The position of the centre of mass is very important for balance. These photographs show sports where balance and centre of mass are very important.

If a sport needs you to stay standing or upright, then a low centre of mass is best. For sports involving jumping, a high centre of mass is better.

Q1 Name three sports where having a low centre of mass would be an advantage.

Q2 When you are on a bike you have a high centre of mass over a small base. Why can this be a problem?

Q3 'Tall people jump the highest.' True or false?
a Discuss the statement given here. Write down as many things as you can that may affect the height that people can jump.
b Can any of your ideas be easily tested or measured? If so, how?

The centre of mass of an object

You are going to find out where an object's centre of mass is. It is possible for the centre of mass to be outside the object.

Apparatus

☐ an irregularly shaped piece of card ☐ shuttlecock ☐ pin ☐ plumbline ☐ table tennis bat

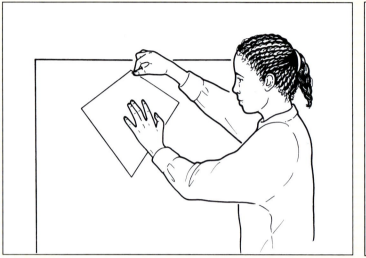

A Pin the piece of card to a notice board. Make sure it can move freely. ▲

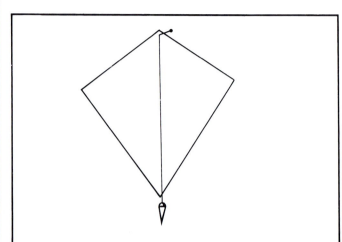

B Next hang the plumbline from the same pin. ▲

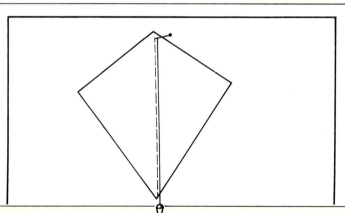

C Mark the line of the plumbline on the card. ▲

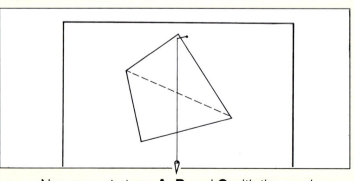

D Now repeat steps **A**, **B** and **C** with the card pinned in a different position. The centre of mass is where the two lines cross. ▲

Q1 What happens when you put the pin through the centre of mass?

Q2 Design an experiment to find the centre of mass of differently shaped people. You could use small people shapes cut out of card. Do not forget to write down how to do the experiment. You will also need a record of your results. Discuss your idea with your teacher. Then try it out.

Q3 Look back at Q3 on page 27. How might you change your answer now?

7 Sports injuries and medicine

Types of injury

Injury is accepted as a **hazard** in most sports. Most injuries affect the muscles and bones. They usually occur in the arms or legs.

Muscle injuries
When too great a force is put on a muscle, the result is a torn muscle.

Tendon injuries
Another injury which is often called a pulled muscle is a damaged tendon. Tendons attach the muscle to the bone. These tendons can be stretched, torn or completely separated. This also happens when too great a force is used. ▼

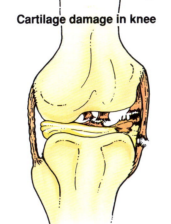

Cartilage damage in knee

Cartilage is found where one bone meets another to form a joint. It protects the ends of the bones. The cartilage can be torn. It may have to be removed. Damage occurs most often in the knee. This happens when the knee is strained. Cartilage damage can cause the knee to lock in place, give way, or click when it is moved. ◄

Dislocations happen when the bones of a joint are forced out of place by a blow. The most common ones are to the shoulder (in sports like rugby, judo, American football). The finger joints are often dislocated in basketball and cricket. ▼

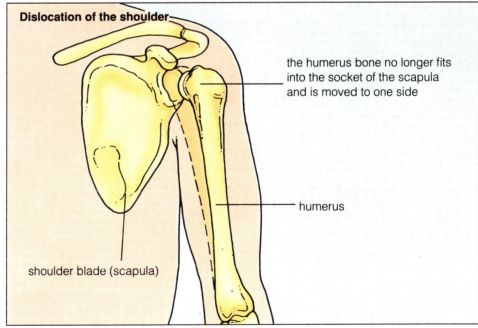

Dislocation of the shoulder

the humerus bone no longer fits into the socket of the scapula and is moved to one side

humerus

shoulder blade (scapula)

Sprains happen when the ligaments are stretched or torn a little bit but not enough to cause a dislocation.

Fractures (broken bones) are usually caused by a direct blow. A stress fracture is a slight crack in a bone. It is caused by a repeated strong force on the bone. Jogging, running and sprinting can all cause stress fractures.

Q1 Which sports are more likely to cause dislocations?

Q2 Explain your answer to Q1.

Q3 Have you ever been injured playing sport? If so, explain how it happened.

Sports injury treatment

As more and more people take up sports, more and more people get injured. Hospital **casualty** staff are not experts on sports injury treatment. There are special doctors and **physiotherapists** who know how to treat sports injuries. However, there are not enough of them.

Sports medicine is not just about injuries. It includes diet, strength, fitness, stamina and suppleness.

⚠️ Many muscle and tendon injuries can be avoided by warming up before playing.

These graphs give information about people treated at a sports injuries clinic. ▼

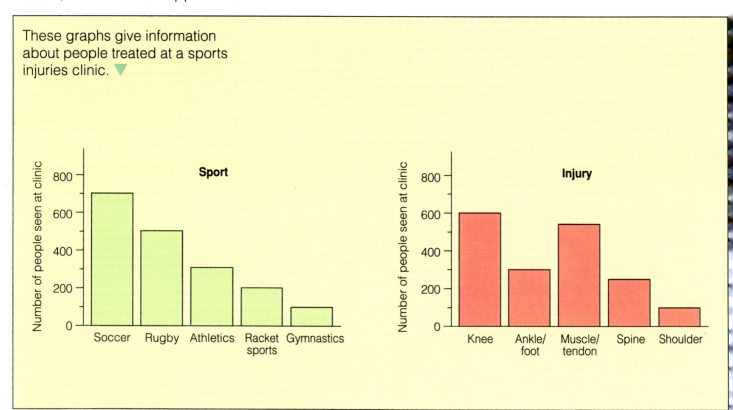

Sports injuries can be prevented by:

☐ being fit for the sport or game you play

☐ obeying the rules

☐ wearing the right clothes and shoes

☐ using common sense

Reckless play is the most common cause of injury

Q1 Why are there more injuries in soccer and rugby than in any other sports?

Q2 Why do you think gymnasts have fewer injuries than people who do the other sports shown in the first graph?

The use of drugs and medicines in sport

The word **drug** can mean many things. **Caffeine** in tea, coffee and fizzy drinks, **nicotine** in cigarette smoke and addictive painkillers like **morphine** and **heroin** are all drugs.

As long ago as the last century, drugs were taken to improve performance and **endurance** in sports. In 1890 a cyclist died from using a stimulant called **ephedrine**. He was using it to keep his body going during a long race. Using drugs or medicines to improve physical or mental performance in sport *artificially* is called **doping**. Refusal to take a dope test leads to a disqualification. Most sports associations, including the Football Association, the Amateur Athletic Association and the Olympic Association, have strict rules on which drugs and medicines can be used.

▲ In 1988 Ben Johnson won the 100m gold medal at the Seoul Olympics. He was disqualified after he was found to have been using drugs to improve his performance.

In 1960 a Danish long-distance runner collapsed and died. He had been using a drug to increase the blood supply to his legs. It killed him.

There are three main types of banned substance:	
Stimulants	These give a sense of well being. Sometimes they allow a person to keep going even when the body is very tired. They include caffeine, **amphetamines** and ephedrine.
Narcotics – painkillers (analgesics)	Help to overcome pain. They include morphine, heroin and codeine.
Anabolic steroids (hormones)	Used to build up and strengthen muscle. They can also improve recovery rates.

◀ ▲ Training, dedication and determination are the qualities of a good athlete.

Q1 Do you think taking drugs to improve performance is cheating? Make a list of points for and against drugs in sport.

8 Sport and leisure

'Sport for all'

Look at this list of sports

angling	gliding	shooting
archery	golf	skating
athletics	gymnastics	skiing
badminton	handball	snooker
ballooning	hockey	softball
baseball	ice hockey	speedway
basketball	judo	squash
BMX	lacrosse	surfing
bobsleigh	marathon	swimming
boxing	running	table tennis
bowling	motocross	tennis
canoeing	mountaineering	volleyball
cricket	netball	water polo
cycling	paddleball	weight lifting
darts	pentathlon	windsurfing
decathlon	polo	wrestling
diving	pool	yachting
equestrianism	rounders	and many more
fencing	rowing	
football	rugby	

Sport on television

'Playing is better than watching ... it'll bring out the best in you ... You're never too old to start, so take up something now.'

In 1972 the Sports Council started the 'Sport for All' campaign. It is to encourage people of all ages and abilities to take part in sport.

As well as taking part in sports, many people watch sport at matches, or on the television. Sports such as snooker, skating and gymnastics have become popular because of television. Some people complain that there is too much sport on television. What do you think?

Q1 Which of the sports listed above are ball games?

Q2 What do bobsleighing, cycling, gliding and motocross have in common?

Q3 Which sports need snow or ice?

Q4 Which ones involve great stamina and endurance?

Q5 Do you play a sport?

Q6 Apart from keeping fit, write down some other reasons for taking part in sport.

Q7 25-year-old Darren plays football for his local pub team. He drinks a few pints of beer every night, has a game of darts and stops for a burger and chips on the way home. He is trying to cut down his smoking. He has recently been dropped from the 1st team to the 2nd team. What advice would you give him to help him get back to the 1st team?

Q8 How many hours of sport will be on television this week? (Use television programme listings to answer this question.)

Q9 Make a list of the sports that will be shown on television.

Extension exercise 6 can be used now.